# Adaptive Project Management:
# Leading
# Complex and Uncertain
# Projects

Andy Silber

Published by BookLocker.com, Inc., St. Petersburg, Florida.

Printed on acid-free paper.

BookLocker.com, Inc.
2017

First Edition

# DISCLAIMER

This book details the author's personal experiences with and opinions about project management and product development.

The author and publisher are providing this book and its contents on an "as is" basis and make no representations or warranties of any kind with respect to this book or its contents. The author and publisher disclaim all such representations and warranties, including for example warranties of merchantability and legal advice for a particular purpose. In addition, the author and publisher do not represent or warrant that the information accessible via this book is accurate, complete or current.

The statements made about products and services have not been evaluated by the U.S. government. Please consult with your own legal, accounting, tax, or other licensed professional regarding the suggestions and recommendations made in this book.

Except as specifically stated in this book, neither the author or publisher, nor any authors, contributors, or other representatives will be liable for damages arising out of or in connection with the use of this book. This is a comprehensive limitation of liability that applies to all damages of any kind, including (without limitation) compensatory; direct, indirect or consequential damages; loss of data, income or profit; loss of or damage to property and claims of third parties.

This book provides content related to project management. As such, use of this book implies your acceptance of this disclaimer.

# Dedication

This book is dedicated to everyone who has struggled in their product development efforts because of poor project management. My hope is that this book will help create more successful products by improving the way these challenging projects are managed.

# Acknowledgments

Many of the ideas in *Adaptive Project Management* were fleshed out in discussions with Ian MacDuff during our time building the project management practice at Product Creation Studio. What I call adaptive project management, Ian calls GPS project management. A GPS (the project manager) tells the driver (the stakeholders) where they are and suggests routes to where they want to go. If there's an accident on the road (e.g., an assumption is discovered to be wrong, invalidating the plan), the GPS can then suggest new routes. The driver always makes the decision, not the GPS.

The section discussing prototypes is taken largely from a presentation given by Scott Thielman of Product Creation Studio.

I'd like to thank everyone I've ever worked with. I've learned so much from all of you. What makes a job great is working with interesting problems and great people, and I've been blessed to be involved with many of both.

Some sections of this book are based on blog posts I wrote and were previously published by Product

Creation Studio, LiquidPlanner or my own blog www.asilberlining.com.

I'd also like to thank the many editors and proofreaders for their contributions to making this book what it is.

# Table Of Contents

## Is This Book for You?

The ideal reader of this book has some knowledge of project management but does not need to be an expert.

Some likely readers include:

- An experienced project manager who has found the methodologies they've used in the past wanting.
- An experienced individual contributor who is being asked to manage their first project, which is full of uncertainty and complexity.
- An individual contributor who is working in an adaptive environment.
- An executive or stakeholder who is frustrated by seeing projects fail for the same reasons again and again and is looking for a better way to manage their efforts.

## About the Author

When people who know me as a project manager hear I have a PhD in astrophysics from MIT, I often get questions like:

- "Do you use what you learned in school?"
- "Do you regret spending all of that time in school?"
- "Do you miss astronomy?"
- "Can you do my star chart?"

The answers are: "More than you would think," "No," "A bit, but I don't regret moving on," and "Not astrology, astronomy."

I've found that physics is a perfect background for project management of hardware product development. I know enough of all of the disciplines to see gaps and recognize risks, but not so much that I'm tempted to actually do the work. When I'm speaking with engineers, I frequently ask technical questions or have suggestions that further their work.

My first job was at Neopath, where I worked on improving an automated microscope that scanned

Pap smear slides to diagnose cervical cancer. From there, I went on to develop industrial equipment, medical devices, and consumer electronics. When I was at Calypso Medical, I started out as an individual contributor with ownership of part of a device that improved radiation therapy for the treatment of cancer. When I delivered my prototype and demonstrated how my preferred solution would meet our requirements, I was tasked with finding and managing an outsider company to implement my concept, beginning my transition to project management. Since then, my career has focused on project management and team leadership for product development.

# Introduction

If you look at the top selling project management books, most can be put into one of three categories:

- Agile (software focused)
- "Standard," or waterfall, project management
- PMP exam preparation

In this book, I'll explain why neither agile nor waterfall project management work well for projects where uncertainty and complexity are both present. In my professional life, I've worked on developing innovative hardware products where we started out with only an idea, some math, or back-of-the-napkin sketches. The goal seemed likely to be obtainable, but it would take teams of engineers across multiple disciplines to invent solutions to the challenges we would come across. These types of projects don't work well with either waterfall (which deals with uncertainty poorly) or agile (which deals with complexity poorly).

But just because the most popular project management paradigms don't work, that's no reason to throw up our hands and walk through the development process randomly. There are tools, many taken from waterfall and agile processes, that reduce risks and lower costs.

This book describes adaptive project management, which is value-added planning, with the awareness that our knowledge is incomplete. As our understanding increases, we update our plan to reflect this knowledge. As you near the end of a project, the planning asymptotically approaches waterfall. The project manager focuses on reducing risk and uncertainty, managing the team so that everyone is working on the right tasks and has the information and resources they need. The project manager also manages communications with the stakeholders (the people accountable for a successful project).

If you use the tools described here for projects with uncertainty and complexity, you're more likely to achieve your goals than if you follow waterfall, agile, or no project management at all. But there's no guarantee that you'll achieve success. For one thing, what you want to do may not be physically or technically possible, or the cost of building the final product will be too high. But even if you fail, there's value in following an adaptive approach: the stakeholders are inside the project, so there will be no surprises, and the failure should happen early in the process because you've focused on high-risk elements at the start. The examples I'll present are from hardware product development, but the ideas here work for any project where complexity and uncertainty are high.

# What Complexity and Uncertainty Mean

The terms *complexity* and *uncertainty* get used a lot in this book, so I want to be clear what I mean. Let's start with what your stakeholder (the *customer* of the project) is asking for:

**Figure 1: What your stakeholder wants. The baby's face with the flower is critical (a requirement). The placement of the lily pads is not critical.**

But what you start with is this:

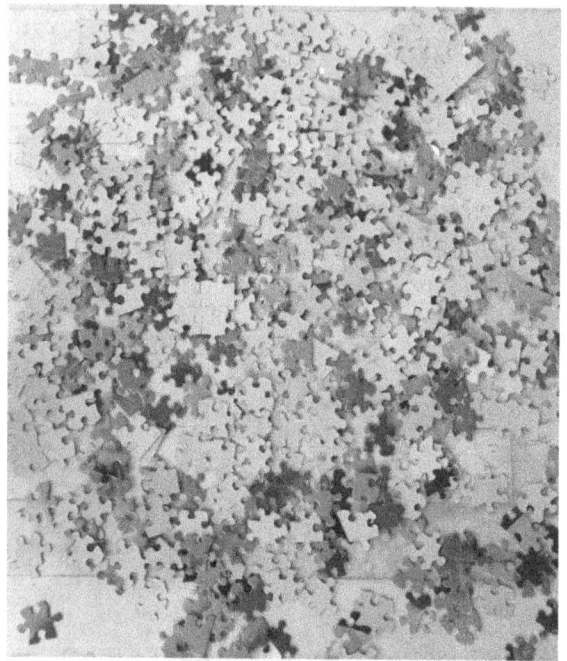

**Figure 2: The project starting point. The upside-down pieces represent uncertainty; the large number of pieces represents complexity.**

In the beginning, you have hundreds of pieces that need to be combined (complexity), and you don't fully understand many of them (uncertainty). Your first step is reducing uncertainty by flipping over the pieces. Your most important question is, "Can I meet my stakeholder's expectation?" As Donald Rumsfeld famously said:

> *There are also unknown unknowns—the ones we don't know we don't know. And if one looks throughout the history of our country and other free countries, it is the latter category that tend to be the difficult ones.*

So we focus on reducing uncertainty, especially the unknown unknowns. This might mean understanding what the end user wants or the limitations of currently available technology.

**Figure 3: The elements of my project after working to reduce uncertainty (i.e., flipping pieces over).**

I still don't know everything, but I do have enough to build my first prototype, the baby's face. This is the requirement my stakeholder cares the most about, so if we can't do this, we should kill the project.

**Figure 4: The first prototype demonstrates that we can successfully build the baby's face with the given resources, meeting the stakeholder's primary requirement.**

Our first prototype shows that we have the resources needed to accomplish the primary requirement, building the baby's face. We show this prototype to the stakeholder and get their agreement that we've met their requirement. It's possible that they don't feel the requirement has been met; maybe the lines between the pieces are unacceptable. If everyone agrees we're on a path that can lead to success, we build a second prototype that includes the flower.

**Figure 5: The second prototype shows a problem. Given the resources at our disposal, we're unable to meet the requirement for the flower.**

Given the pieces of the puzzle we have, we can't complete the flower. This leads to a discussion with the stakeholder to determine if we:

1. Continue on, relaxing the requirement such that our current solution is acceptable.
2. Increase resources so that we can make a custom part from scratch to allow us to meet the original requirement.
3. Kill the project.

Let's continue on with option 1. We now complete the project given our original resources, but with relaxed requirements.

Figure 6: The completed project. As is often the case, there are a few pieces missing from the original concept and reality is not as pretty as the artistic vision that started the project.

We now have a completed project. Along the way, the stakeholders saw the results from the prototypes, so they're not surprised by the final product.

Unlike assembling a jigsaw puzzle, in product development, you start with the hard parts and leave the easy parts (e.g., the edges) for the end. We started with the baby's face because it was critical, not because it was easy.

# Why Waterfall Project Management Isn't Well Suited for Problems with Lots of Uncertainty

Waterfall project management is the paradigm taught by the Project Management Institute, the grantee of the Project Management Professional (PMP) certification. It's so ubiquitous that if someone says "project management," waterfall is probably what they mean. It's an incredibly powerful approach, allowing one to manage immensely complex projects efficiently and successfully. The intention of waterfall is that the tasks flow from one to another as effortlessly and inevitably as water falling down a mountain.

This approach is well suited for highly complex, yet predictable, projects like constructing skyscrapers or organizing a national tour of the Rolling Stones. Just as you would never start building a skyscraper without a complete design or begin organizing a tour without a band, you also need a complete plan before beginning a waterfall project. In waterfall, every task is listed, and the amount of effort it will take is estimated: this is called the work breakdown structure (WBS). Project managers note the order in which tasks must be carried out, and specific tasks are assigned to the responsible parties. Tools like Microsoft Project are used to turn these inputs into a schedule or a Gantt chart. As you move through the plan, materials are delivered to the right place at the right time. If they arrive too early, there is no place to store them; arrive too late, and it delays the project. Some of these inputs, like booking a large venue for a concert, need to be scheduled far in advance. One of the advantages of waterfall is its ability to determine the critical path, the tasks that if delayed by a day will mean the final completion date of the project is also delayed by a day. As a project manager, this helps you know where to focus your resources and attention.

The types of projects that are best suited for waterfall are those that exhibit little change in the needed skills and tools from one project to the next. The vast majority of the workers, from people swinging hammers on a construction project to people setting up the speakers for a concert, will have worked on very similar projects using identical tools. This routineness vastly reduces the uncertainty and risk. It also makes estimating how long tasks will take extremely accurate since the practitioners have done these tasks before.

This is a far different situation than in product development. Tools like miniature sensors, 3-D printers, and low-power microprocessors are constantly changing. This makes what was impossible a few years ago achievable today. While on the one hand, these innovations are incredibly exciting, they also make the product development process far more uncertain and risky; it becomes impossible to assemble a precise project plan. If you have a patent attorney on speed dial, you are not going to be able to build a credible Gantt chart.

For example, I once worked on a product with a system-on-a-chip (SOC) processor with Bluetooth low energy (BLE) built in. As we started pushing the envelope on how many devices this product would communicate with, we discovered that the SOC wasn't able to do what we needed. The supplier of the chip didn't even know the limitations because these functions had never been tried before.

Before a waterfall project is launched, every effort must be made to reduce uncertainty (e.g., soil samples taken and computer models run) because waterfall does a bad job of dealing with changes once the project is launched. When the unexpected happens (e.g., a big storm the night of a concert), these projects tend to fail spectacularly. In product development, the unexpected is typical, so we need a development process that bends like a tree during a storm, not one that shatters like a window.

The development of the F-35 fighter jet is a good example of the misuse of waterfall project management. This program created variants of a fighter aircraft suitable for the US Air Force, Marines, Navy, and the militaries of at least ten other countries. The requirements for this program were incredibly challenging: the plane needed to have stealth capabilities, be able to launch ground attacks, maintain air superiority, and have the ability to take off and land vertically. The plan was to design, test, and start production concurrently. Most of the people working on the project had little or no experience designing a fighter from scratch, and the experience needed to build a solid plan didn't exist. Since they were using the latest technology, nothing in the project was routine. Testing revealed problems on planes that had already been manufactured, resulting in rework for the planes already built. The

F-35 is plagued with design flaws and is at least $165 **billion** over budget.

In my own experience, I worked at one company that required a detailed waterfall plan for the entirety of a project—including commitments from managers on three different continents that engineers would be available on the exact days the plan said they'd be needed. Since the first part of the project was conducting feasibility studies and developing the architecture of the solution, it was impossible to accurately say how long it was going to take and when the engineers were going to be needed. One of the executives suggested we do statistical modeling to estimate how long tasks were going to take, not understanding that these tasks were poorly defined and had never been done before. We estimated durations based upon the time taken on similar projects, but great uncertainty remained. After failing to launch the project and restarting several times with new project managers, the company finally gave up. If they had followed an adaptive process, they would still have failed since the concept was technically unfeasible, but it would have happened sooner, saving money and freeing the team to consider better ways of meeting the market need.

Waterfall project management has many useful tools for projects with uncertainty and complexity, like establishing dependencies and a critical path. However, the paradigm of creating a complete plan at the beginning of a project is based on the incorrect assumption that we know enough in the beginning to create an accurate plan all the way through to product launch.

# Agile is Too Flexible for Complex Projects

"The Agile Manifesto," published in 2001, describes what the authors felt was most important when developing software:

***Manifesto for Agile Software Development**

*We are uncovering better ways of developing software by doing it and helping others do it. Through this work, we have come to value:*

- *Individuals and interactions over processes and tools*

- *Working software over comprehensive documentation*

- *Customer collaboration over contract negotiation*

- *Responding to change over following a plan*

*That is, while there is value in the items on the right, we value the items on the left more.*

When I read the manifesto, I'm struck by how reasonable the statements are in the context of product development and adaptive project management, especially that responding to change is more important than following a plan. What's the point of following a plan that's based on outdated knowledge or bad assumptions?

The agile methodology is a project management approach that has grown from this manifesto. In agile, the features the customer wants are listed, often on sticky notes attached to a board in the common workspace. The work is broken into groups that can be implemented in a short amount of time (usually a few weeks), called sprints. At the end of each sprint, a working and tested product is shown to the product owner (i.e., someone who speaks for the interests of the customer) who then gives feedback. This feedback is used to decide what work will be incorporated into the next sprint.

Agile is unlike waterfall in almost every way imaginable. In waterfall, one needs requirements, a design, and a plan before you start executing. In agile, all you need to start writing code is a product owner, some developers, a vague definition of the problem you're solving, and an idea of how you might solve that problem. Requirements might never be gathered together in a document. A product design isn't created, but evolves. The plan is just a list of the features that will be implemented, with a priority ranking, and this list will change at the end of each sprint. Agile focuses on executing short development cycles, creating value, and generating lots of customer feedback. In agile development, the creation of detailed requirements, documentation, and plans are seen as a waste of time because these artifacts constantly change and don't create value for the customer.

Agile works well when developing a web site or software-as-a-service, where the time to release an iteration is short and the cost of a bad decision is low. These are the types of problems that agile was created to solve. When designing hardware products, agile's focus on customer input, early and frequent testing, and adherence to creating value are all improvements over the waterfall method of project management.

One of agile's strengths is the focus on testing. As code is developed, tests are written that demonstrate the code does what's expected. Anytime a change is made in the code, automated test tools are run showing that the entire program still works as intended. These tests are often the best form of documentation because if they are out of sync with the code, the test will fail. This focus on testing catches errors early, when they are still easy to fix.

The most important lesson we take from agile is the relationship with the stakeholder. In agile, you're constantly getting feedback from a stakeholder, who represents the end user. Showing early work to your stakeholder allows for misunderstandings or uncertainties in the requirements to be quickly discovered and remedied, like the lines in the jigsaw puzzle from the example above. In adaptive project management, it's critical that the stakeholder has this tight level of involvement.

What agile does poorly is deal with technical risks, tasks that must be done in a particular order, and long lead times. Imagine building a skyscraper with an agile approach: you're experimenting with a new concrete formulation, and it takes longer to dry than expected; the electrician shows up two weeks after the walls were closed up; the client looks at the half-constructed building and requests an additional floor for a gym he just decided he wants.

Rapid prototyping tools like 3-D printing have made it easier to bring some of the agile philosophy to hardware product development, but there's still a fundamental difference in the way hardware and software are developed. With the automated testing and light documentation requirements preferred by agile, it takes about an hour to release a new iteration of software. An iteration of a hardware product can easily take two months, not including the time needed to change the design, which may include altering or building manufacturing tools, obtaining all of the parts needed, building and populating new printed circuit boards, scheduling time on a production line, and building new test fixtures.

The agile approach allows one to start developing very quickly but at the risk of heading down a dead-end and having to start over, which increases project costs and the time to complete. As described in the classic novel *Zen and the Art of Motorcycle Maintenance* by Robert M. Pirsig, having to rework sucks one's gumption (the energy needed to do high-quality work). It is better to move forward thoughtfully than need to repeat work because you left a gasket out in an early step of rebuilding a carburetor.

In developing a hardware product, it is important to have a destination (requirements) and a map (a design and a plan) before one begins a journey. For agile, the destination is vague, and the map is nonexistent. This is acceptable for software development, where the cost of making a mistake is minimal. In hardware development, mistakes can be expensive in terms of time and money; one needs a process that reduces both the chance of making mistakes and the cost of the ones you will inevitably make.

## Adaptive Project Management: The Right Balance Between Structure and Flexibility

So, if agile and waterfall methodologies work poorly for projects like developing a hardware product, what's a PM to do? While a strict adherence to these paradigms will lead to agony and pain for projects where there's complexity and uncertainty, there is still much of value in both approaches. Waterfall and agile can be combined into an effective project management paradigm. I call this approach adaptive project management.

Helmuth Von Moltke said that "No plan survives contact with the enemy." Adaptive project management is based on this fact. Just because the plan will change doesn't mean that it had no value, but knowing this, you should restrict your effort to the part of planning that adds value.

This approach manages risks and tracks the critical path while recognizing that our knowledge is inherently incomplete and replanning will be necessary. Using an adaptive approach, the project manager does just enough waterfall-style planning to be confident that the team is working on the right tasks. These tasks are chosen to minimize project risk as early as possible while making sure that long lead-time tasks are completed when they need to be. As the project moves forward, the plan is reworked and expanded to leverage the most recent understanding of the project. This combined approach is similar to what the Project Management Institute calls rolling wave planning and others call hybrid project management.

Writer E. L. Doctorow described writing as being "like driving a car at night: You never see further than your headlights, but you can make the whole trip that way." This metaphor is just as apt for product development. A good map is helpful, but it doesn't absolve the driver from paying attention. The map won't mention the cow in the road, the traffic jam, or the bypass route that was built since it was drawn. Your map will be imperfect, but if you keep moving toward your destination, you might get there. Or you might discover the road is impassable.

The product development process is a march from high uncertainty and risk toward a de-risked, highly specified product that's capable of being manufactured and sold. Early development phases will focus on understanding customer needs, the competitive landscape of the market, and technological opportunities. Middle phases will focus on proof-of-concept experiments and narrowing your possible concepts to those that best solve the problem at hand. Late phases will focus on designing, building, and testing commercial devices. The final phase will be transferring a manufacturable design to production. As you remove uncertainty, your plan should look more and more like a waterfall plan.

The tools for adaptive project management are similar to those for waterfall. Project managers create a list of tasks that need to be completed (the WBS), but only build a detailed schedule for the tasks that are in the near future, have long lead times, or have high risks. The WBS is reviewed often, and tasks are added as they are discovered. Like an agile approach, the planning is constantly focused on adding value. The difference is that there's an acknowledgment that some high-value tasks don't show value immediately. One must do enough planning to understand what those tasks are and when they need to begin.

I was part of a team that used this process to develop a consumer electronics product. Reliability was a primary concern, so for each prototype iteration, we built enough samples to demonstrate not only that the product would work under normal conditions, but that it would continue to work after being heated, cooled, pulled, twisted, soaked in water, and coated in sweat. We were constantly updating our design based on the results of testing, but to meet our schedule, we started building the next round of prototypes before the testing on the previous variant was complete. To stay on target, we needed a detailed waterfall schedule for each build, but what was in the build was unknown until the last minute, when the design was released. At any given moment, the design and the plan were based on the best available information. The program met our schedule and the product was reliable and of high quality: it was a design the team was proud of.

I think the most interesting example of a challenging project that used an adaptive approach was Lewis and Clark's exploration of North America. They started out with a mission statement from their primary stakeholder, President Jefferson:

*The object of your mission is to explore the Missouri River, & such principle stream of it, as, by it's [sic] course & communication with the waters of the Pacific ocean, may offer the most direct & practicable water communication across this continent, for the purpose of commerce.*

**Figure 7: Lewis and Clark's Corps of Discovery at Three Forks**

They had a literal map that got them started up the Missouri River and a figurative map that informed them of some of the risks and challenges they would face. They selected members for the Corps of Discovery with the skills they would need, like speaking the languages of the people they would meet, boat building, and hunting. They gathered gifts for the tribes they would meet and supplies like guns, boats, food, and scientific instruments. Selecting and training the Corps of Discovery and preparing for the voyage took about a year. As they moved west, they drew the map that others would follow. Every encounter with a friendly tribe added to their knowledge and increased their chances of success. They added team members as they moved west as well, including Sacagawea and her husband, who joined the expedition six months after the journey began. If they had used a waterfall approach, Lewis and Clark would still be in St. Louis building a Gantt chart. Using an agile approach would have been worse; they would have left without sufficient planning and preparation, possibly ending in the death of the team and the failure of their mission. Only an adaptive approach had a reasonable chance of success, and even then, success was far from assured.

Though adaptive project management is not optimal for all projects, it works better than waterfall when there are more than a handful of unknowns. It works better than agile when early mistakes can doom a project or when the interactions between tasks are complex. But even the best possible project management approach will not make the impossible possible; it just reduces the cost of learning that lesson. You may come across a waterfall that you didn't expect, but that doesn't mean you have to go over it and drown.

# Managing a Project Means Managing a Team

Some think that the most important contribution of a project manager (PM) is building the plan: taking the work that needs to be done and breaking it down into tasks, identifying the dependencies, feeding those tasks into a project management tool, working with the functional managers to build the right team, and assigning tasks to your team members. From there, the team implements the plan, and the PM monitors the work to make sure the team is on schedule. This approach works for projects with few unknowns and no need to innovate.

But this approach to project management doesn't work if your team needs to be inventive. Creating a safe space for the team to innovate is critical in adaptive project management. If everyone is in a rush to get something ready for the next deadline, the creative juices won't be flowing. Your team members need time to take a walk, chat, or do whatever it is that gets them into their creative space. And no one will be willing to try an approach that might fail if failure is not an option.

## Traits of an Adaptive PM

People often think of a PM either as someone who simply checks boxes or as a strong leader who micromanages every individual contributor, but neither of those approaches works well for an adaptive project. The tasks are often too ill-defined for the PM to just check boxes, and no one likes working with a micromanager.

The PM does not need to understand the technical skills well enough to perform the tasks, only well enough to have meaningful conversations with the individual contributors (ICs) about their uncertainties and risks. It's a good sign if you're occasionally asking questions that make the IC pause and think. Your role is not that of a product architect, the person who is driving the technical decisions, although sometimes, for small projects, the PM and the architect are the same person.

Building a project plan is a bit like solving a puzzle: bringing the clues/data together and getting it to fit like a game of *Tetris*. You need to enjoy that aspect of the job because, in an adaptive project, you'll be replanning regularly.

You need to enjoy helping people do their best work. Your job is to be a force multiplier; the team is doing better work because of the leadership you provide. Typically, they get to do the fun stuff (e.g., build and test prototypes) and you get to fill out status reports, update schedules, and deal with cranky stakeholders.

## Service Leadership

The PM's role is one of service leadership, making sure barriers are removed, information is flowing where it's needed, and resources are sufficient for the tasks at hand. It's not the PM's job to make the decisions, just to make sure the decisions are being made by the correct people who have the best available information. The team should see the PM as someone to go to for help, regardless of what help they need.

One of the PM's most important roles is to protect the team from crazy stakeholders and unreasonable expectations. If you see someone getting set up for failure, you stand in front of that train and stop it in its track. Otherwise, the failure will disrupt the project and demoralize the team. If something goes wrong, the PM takes the heat, representing the team, and never throws someone under the bus.

## Know What Your Team Is Doing Without Getting in Their Way

When creating a plan for building a skyscraper, an experienced construction project manager will know how long it takes to perform routine tasks like pouring concrete or wiring the lights and be able to build the plan without talking to the people who will actually do the work. But when innovation is involved, few tasks are routine, so the PM must work with the entire team to create the plan.

Once the project is launched, the PM needs to engage the team daily, possibly through scrum-style stand-up meetings, so that the plan can be updated and the stakeholders informed of changes as the plan evolves. At any moment, a stakeholder should be able to call the PM and get status on the biggest risks, the latest learnings, how the plan has changed, and progress toward your next deliverable. Regular reports must be sent to the stakeholders describing the status of the project, especially any major lessons learned or departures from expectations.

## Focus on Solving Problems, Not Completing Tasks

One way the PM builds a culture of innovation is to focus on *problems to solve* as the core activity, as opposed to *tasks to perform*. This gives the team the flexibility to go where their expertise and discoveries lead them. This, in turn, results in better outcomes and higher morale.

I once worked with a client who assigned us a *task* to measure the electronic noise of a prototype, but the *question they were trying to answer* was whether shielding would be needed. The team discovered that other parts of the system were designed in a way that the answer was obvious; the noise would be unacceptably high without shielding. Our suggestion was to improve the filtering before we measured the noise. If the team had just performed the task, we would have wasted a lot of time and money and not gotten any closer to a solution.

## Failure Is Not Only an Option; It's a Requirement

If your team feels that failure is not an option, it's unlikely they will take the necessary risks in order to find an innovative solution. Your stakeholders may be concerned with wasted time and effort if a concept doesn't pan out, but an effort is only wasted if you learned nothing from it. If you spent three days building a prototype that you should have known wouldn't work based on reading a Wikipedia article, that's wasted effort. But if the reason the concept didn't work was not obvious without spending the three days building and testing the prototype, then it was a success.

As PM, it's your job to build a plan that doesn't assume the project follows the "happy path" where everything goes right. If you do, then ideas that don't pan out lead to delay and disappointment. This creates stress on the stakeholders and the team, leading to a future of risk aversion. It's hard to build "failure" into your plan because your stakeholders want you to move fast, but it's even harder to build invention into your plans.

## Be Part of Something Vital

It's critical that everyone on the team feels protected so they will have the confidence to innovate. Learning and intelligent risk-taking must always be rewarded, even if all you learn is that the project isn't feasible.

In an innovative culture, the distinction between the people manager and the project manager becomes blurred. The PM can't treat the practitioners as cogs. Your team members will often be crossing disciplines to invent their way out of a problem. This requires much more communication and trust. It's your job as the PM to nurture a team where this can happen.

# Building a Schedule

One of the advantages of adaptive over waterfall is that once the project has a green-light, we can start doing risk-reduction or other development work while we're building our plan. In adaptive project management, we de-emphasize building a schedule, and it's usually clear what some of the highest priority work is before the plan is complete. But we still build a plan.

It's critical that expectations are reasonable for accuracy on the level of effort required for tasks that have never been done before. A way to describe this is to ask the primary stakeholder how long the drive takes from work to home. They might say something like, "Around thirty minutes plus or minus fifteen minutes, depending on time of day, weather, if there's a game or big concert." So, if they can only estimate to 50 percent precision something they do five days a week, how can they expect you to more accurately estimate something no one has ever done?

I once had a head of project management try to teach me how to build a schedule for a project with lots of uncertainty. His experience was from a factory, so he suggested estimating the time to complete a task using the median of a sample of similar tasks. Since most of the tasks in the work breakdown structure were poorly understood and had never been performed before, this was not helpful. This project was full of invention, and the engineers who had looked at the problem gave it a 30 percent chance of success (spoiler alert: after trying for several years, the project failed). Yet management wanted a schedule with high precision, and a slip by as much as a week after nine months would lead to the PM being yelled at by the CEO. This was an environment that crushed innovation.

That doesn't mean you don't build a waterfall-style schedule; it just means that everyone should understand the assumptions and uncertainties that went into it. One of the reasons it's worth the effort to build the schedule is to see the critical path, which is the set of tasks that, if delayed by one day, will result in the deliverable also being delayed by one day. Each deliverable can have its own critical path. By understanding what tasks are on the critical path, you can understand where to put your focus as PM. The schedule also lets you know if your deliverables might be completed in time to meet your stakeholder's expectations. If the schedule says they won't be available, you can look at adding resources, de-scoping, or other ways to accelerate the project.

Here's an outline of how to build the schedule:

- Create a work breakdown structure.
    - Break the tasks that are coming up soon into smaller chunks (two to five days).
    - Tasks later in the project can remain large.
    - Note any dependencies (tasks that need to be completed before other tasks can begin).
- Estimate the level of effort for each task.
    - Ideally, the estimate is made by the people who will be doing the work. If not, they should be people capable of doing the work. In other words, probably not the PM.
    - Avoid the impulse to do top-down planning (e.g., "We need it by December 3, so that's how long you have").

- o Tell your inner optimist to take a nap. Have your inner realist and inner pessimist work it out. One way to do this is to create optimistic, realistic, and pessimistic estimates. Then ignore the optimistic estimate.
- Build your schedule.
  - o Double-check the duration and dependencies of tasks on the critical path.
  - o Is there any way to do some of these tasks in parallel or move them off the critical path?
  - o You'll need to balance tasks that reduce risk and uncertainty with those that get deliverables completed. When in doubt, focus on reducing risk and uncertainty.
- If deliverables won't be completed when needed, look for ways to accelerate the schedule.
  - o Can you start some tasks early even though this risks wasted work?
  - o Can you add resources and do some tasks in parallel rather than serially?

> Remember that nine women can't make a baby in one month (i.e., added resources don't always speed things up).
> o Is there scope that can be delayed, allowing the deliverable to be ready in time?

By the time the schedule is complete and approved by both the team and the stakeholders, it will be time to update it. With every status report, the schedule should be reviewed as follows:

- Are we on schedule?
  - o If not, why?
- Where has new information allowed the team to make the schedule more realistic?

## Staying on Schedule

Perfection is the enemy of good enough. It's your job as PM to make sure everyone knows what good enough is.

I've found the most important tool for staying on schedule is time boxing (i.e., telling the person doing the work how long they have to do it). Engineers will keep working on a task, making the deliverable better, for as much time as you give them. Instead, make clear the expectation on the quality of the deliverable and the time they have to complete it. At the end of the allotted time, meet and discuss if the output they have at that time is good enough. Often, they'll say, "Just give me a couple more hours and I can make it better." Fight the urge to make it better if it's already good enough; the schedule is probably more important.

## Project Management: More Than Creating a Schedule

Recently, a friend asked me for some project management help. She runs a nonprofit, and their annual fund-raising dinner is a major effort. As a certified project manager, I immediately started thinking about work breakdown structures, Gantt charts, and software that was inexpensive and easy to use. But as we talked and I understood her pain points, it was clear that the path I was heading down would have increased her work without solving any problems. So often, when people, even project managers, think about project management, they think about tasks and schedules. But there's a lot more to managing projects than tracking tasks.

### Procedures

For my friend, I suggested she write down how to accomplish all of the tasks (e.g., on-board a sponsor), one page per task, and gather up all of these procedures in a binder. That way, she could easily delegate tasks to a new volunteer while keeping the quality high. Next year, she'll have the binder to remind her what she did and to pass down to her replacement when the time comes.

Good procedures can mitigate all kinds of risks. For instance, it's surprisingly common for a company to fire/lay off an employee for perfectly legitimate reasons but end up having to settle a legal case because of their lack of a termination procedure. If they would have had a policy for layoffs and firings and followed it, they would have significantly reduced their risk of a lawsuit.

*The Checklist Manifesto* by Atul Gawande is a great book that talks about the amazing value simple procedures can bring to critical tasks like surgery. Having clear, written procedures, which might be just a checklist, helps a team accomplish tasks consistently. You also need to have a straightforward process to update your checklists as the situation changes.

## Risk Register

I've often come across confusion between a risk and an issue. An issue is something you *do* need to deal with, while a risk is something you *might* need to deal with. Your schedule should factor in the effort to resolve all of your issues. Risks complicate planning; you don't typically include solving them in the plan, which means your plan slips when a risk becomes an issue.

The first step is to create a list, or register, of all of the potential risks. Hold a brainstorming session and write down all of the risks, no matter how unlikely or inconsequential. For each risk, rank the probability of its becoming an issue (Prob). Don't worry about getting precise. I suggest a 1–5 ranking, where 1 is very unlikely, and 5 is fairly likely. Next, add severity (Sev), where 1 is extra work, but not enough to delay the project, and 5 is so bad that you'd need to cancel the project. I also like to include a column for importance (Import), which is the result of probability multiplied by severity. The risks should be ranked from high to low importance.

Now, list the mitigations, what the team can do if the risk becomes an issue. After discussing the mitigations, you should reassess the severity. If you can't think of a mitigation, then the severity should be set to 5.

| Risk | Prob | Sev | Import | Mitigation |
|------|------|-----|--------|------------|
| Unit fails electro-compatibility testing | 4 | 3 | 12 | Build Faraday cage around electronics |
| Passive cooling isn't sufficient | 3 | 2 | 6 | Add a fan |

Table 1: This is a simple risk register. More complex versions include things like discoverability (how likely it is the risk will happen, but you won't be able to detect the failure) and contingency (what you will do if the mitigation doesn't solve the problem). I find a simple risk register sufficient and easier to keep current, which is more important.

Just like the other elements of the plan, this is a living document that should be revisited regularly and shared with the stakeholders. For brevity, you can eliminate the risks that have low importance from the version you share with your stakeholders. If a risk has a high probability, I recommend including the mitigation effort in your plan.

## Manage Scope

Many a project has been launched with ambiguous goals and deliverables (a.k.a. scope). It's up to the PM to make sure the work carried out by the team and the deliverables created are in-line with the desires of the stakeholders. This can be especially difficult if the stakeholders are unsure of what they want or are not aligned with each other.

Projects failing because of an undisciplined increase in scope is so common, there's a name for it: scope creep. It's up to the project manager to make sure that changes in scope are handled in a disciplined way, such as with the following chain of events:

1. The stakeholder asks for additional features.
2. The PM leads the team in determining the level of effort needed to accomplish the new work and its impact on the schedule.
3. The PM works with the stakeholder to decide if the new scope is added to the project and if so, at what priority.
4. All planning documents (e.g., requirements, risk register, schedule) are updated if the scope is changed.

Following these steps for even small changes is important for two reasons:

- If your requirements document is out of date, then your testing, which should be driven by the requirements document, will also be out of date.

- By creating a bit of friction in the process, stakeholders are more likely to ask themselves if the new feature is worth the distraction before spinning up the team.

## Communication with Your Stakeholders

In a standard waterfall project, communication can be as simple as "We're on schedule." But in adaptive project management, your plan is incomplete, so constant communication is critical, more in-line with an agile project. Especially early in the program, risk probabilities and severities will be changing as you reduce uncertainty; the schedule will become more fully fleshed out, and options that you might not have been aware of will become apparent. All of this needs to be shared with your stakeholders.

Here, I'm going to say something that's obvious but not as common as we'd all like to think—always be honest, to a fault. I'm not talking about the "dog ate my homework" kind of dishonesty; that's easy to avoid. I'm talking about **optimism**. For instance, you're behind schedule, but the team has a plan to quickly catch up. Let the stakeholder know you're behind schedule, how it happened, and your plan to catch up. When you discover bad news, it's OK to wait a day to figure out a mitigation plan, but if you don't have one by then, share the bad news and let them know you're working on a mitigation plan. As PM, being the truth teller is part of your job.

The most important communication tool is the status report. You should have a standard format so your reader knows where to look to find what they want. Assume that only the first half of the first page is read, so make sure the headlines capture the important facts. Interested parties can read the full report to get details if they want. Your current risk registry, changes in scope, and anything you've learned should be included in the report. The reports should come at a regular cadence of every week or two.

The status reports are important for creating historical documentation of the project. I once managed a project where a stakeholder complained about the scope increase. We went through the status reports that explained in detail each increase in scope over the last several months, which let him know that the time to complain was when the scope changed, not months later.

# Minimally Viable Product and Prototypes

There's some ambiguity on how the terms minimally viable product (MVP) and prototype are used. As used here, an MVP is the product with the fewest features that you believe people will buy given expected market conditions. A prototype is any device, model, or artifact that will aid the developing team in answering one or more questions.

## Minimally Viable Product

An example of an MVP is the 2010 Nissan Leaf electric car. It was the first mass-market electric car sold in the United States outside of California. It had a range of seventy-one miles, making it unattractive for many drivers but still a viable product for many others (the average car is driven less than forty miles a day). Due to the environmental and operating cost advantages (the Leaf has a cost per mile of $0.035, compared to $0.12 for a Toyota Corolla or $0.086 for a Prius), these cars sold reasonably well.

The advantage to releasing an MVP is that you get to learn about how your customers interact with your innovative product in the real world while earning revenue. The risk is you might have misjudged what "minimal" is and released an Apple Newton or Microsoft Bob. Both concepts were later successfully launched as the Palm Pilot/iPhone and Siri/Cortana: the product was a good idea, it's just that the early versions weren't good enough to be viable.

When creating your plan, I would suggest aiming for an MVP+. It's much better to get to market later with features you didn't absolutely need than get to market earlier without features you did need (think the Newton vs. the iPhone). Getting to market first isn't as important as entering the market with something that meets your customers' needs, as Apple has shown repeatedly. They were not the first to have a graphical user interface (Xerox was), a solid-state music player (the Rio from Diamond Multimedia was), or the smartphone (IBM had one fifteen years before the iPhone). The other reason to aim for an MVP+ is that if you have budget and schedule issues, you may need to shed some features in order to stay within your means. If every feature is a must-have, you'll have no options.

## Prototypes

Prototypes live on a continuum of three factors of fidelity as compared to your final product:

- Acts like
- Looks like
- Built like

In general, as the project moves forward, the fidelity of prototypes increases, but your prototypes should have no more fidelity than is needed to answer the question it's built to answer.

There are two things every good prototype must have:

1. Question(s) that the prototype will answer.
2. An audience for the answers.

Different audiences will be swayed by different evidence, so it's important to understand what your audience needs in order to sense if the answer is sufficient. If your audience is an internal stakeholder for a stage-gate review, the level of proof needed is different than if the audience is the FCC or the FDA asking if the product is ready for sale. The sign of a bad prototype is if after using it, the answer to your question is "It depends" or "Maybe."

One of the most common uses of prototypes is to de-risk your project. Look at the highest risk items in your risk register and see if you can reduce either the severity or the probability of those items through the use of a prototype.

In product development, early prototypes might be merely a simulation in software, something made out of Styrofoam and cloth, or something drawn by hand on a piece of paper, and each prototype answers only one question. Later in your project, you might see 3-D printed and machined parts, and the prototypes can be used to answer multiple questions. Toward the end, injection-molded parts might be in the mix, and the prototype starts looking and acting more like your final product.

| Question | Audience | Looks Like | Acts Like | Built Like |
|----------|----------|-----------|-----------|-----------|
| Will left-handed people like the shape of our handheld device? | User experience leadership | High | No | No |
| Does our device meet requirements for electromagnetic compatibility (EMC)? | FCC | No | High | Moderate |
| Will our device survive being dropped from a height of one meter? | Stage-gate review | Moderate | High | High |

**Table 2: Examples of prototype questions, audiences, and fidelity.**

## A Case Study in Determining Requirements: Handing out Free Shoes

I volunteered at the Seattle/King County clinic, an event where volunteers provided over a thousand people per day medical services for free. You could have dentures made, get eyeglasses, and receive vaccinations. There were also people to help navigate the health insurance system, which is no easy task.

Everyone who attended was entitled to a free pair of shoes donated by Brooks. I spent an eight-hour shift helping hand out these shoes.

### The Challenge of Determining Requirements

Imagine picking out a pair of shoes for someone. Then imagine this will be their only shoes for a year, so they really care. Next, imagine that they don't speak English. Add in that they have a mouth full of cotton because they just had a tooth pulled. Imagine that you have a line of people out the door looking to get shoes, and the faster you go, the more people will go home with new shoes. How do you quickly figure out the requirements for the right pair of shoes for this client?

In finding shoes, I did a few things:

- Asked their shoe size.
- Asked what colors they liked.
- Looked at their current shoes and clothing to guess if they'd prefer flashy or subtle styles.

Based on my inferences, I would pick out three pairs of shoes of varying styles and see which pair they liked best. Hopefully, they would like one of the pairs, and if the shoes didn't fit, then I'd have a better idea of what they'd like for round two. It was common for someone to say they liked blue but then pick the one green pair I brought back, so you can't just go by what someone says.

One of the hardest tasks of a PM is to extract requirements from the stakeholders. Often you speak different languages (e.g., marketing vs. mechanical engineering) and the stakeholder doesn't completely understand what they want, but they're in a rush to get it (whatever it is).

Not having clear requirements creates a risk that the implementation team will head off in the wrong direction, so it's worth the time to get it right at the start. If there's ambiguity, make it explicit (e.g., we want the mass to be small since it's a handheld device, but we're not exactly sure how heavy is too heavy).

Giving your stakeholders choices is a good way to understand their requirements (e.g., would you prefer having something that weighs one kilogram and is available in a month or 600 grams and is available in two months?). When working with stakeholders, it's fine to infer what people want based on whatever clues you have in order to get the process started, but it's important to understand the difference between implication (what people mean but don't say) and inference (what you think people mean that they aren't saying). If you make an inference, it's important to ask if what you're inferring is what they're implying.

## It's a Challenge Matching Vague Requirements to Constrained Solutions

An older woman told me that she likes black and white and she wears size 8 1/2 shoes. I was faced with this:

My solution is in that box. It's not going to be exactly what she wants. It might not even be close. It's going to be the best I can do given my constraints of time and the shoes in that box.

Often, stakeholders want something that's just not going to happen. Maybe they can't afford it, they're constrained by schedule, or it's just not physically possible. You have to separate out what the stakeholder *wants* from what they *need*. They *need* a pair of shoes to protect their feet; they *want* something that matches their personal style. In my case, most of my clients had reasonable expectations. That's not always the case.

# Software Tools

There are countless tools that can be used for project management, and I'm not even going to attempt to review them. My goal here is to describe the types of tools you might use and what to look for from the perspective of adaptive project management.

## Bug Tracking

One must-have is bug tracking. In this context, a bug can be a problem that needs to be fixed or an idea for an improvement. If you find a bug that you can't resolve immediately, you need a place to track it. It's acceptable to ship a product with known bugs that you've decided are low enough risk. What's unforgivable is to ship with critical bugs that you've just lost track of.

Every bug should be logged with a short description, how to replicate it, the current stage in the effort to fix it (e.g., on hold, active, testing, resolved), and priority. This is a bare minimum.

It is possible to just use a spreadsheet that's stored in the cloud, like Google Sheets. Do not use a non-cloud-based spreadsheet like an Excel file; it's only a matter of time before there are two copies and no one is sure which is the "right" copy, probably because neither one is current. If you must use Excel, make sure you use a tool like SharePoint to manage the official version.

A much better solution and one that's not much more difficult is to use is a database like Bugzilla or Trac. These tools allow a much richer description of the bug and integrate with some software development tools. If your organization is already using a bug tracking tool, that's probably the best one to use.

## Communication

The most common communication tool is e-mail, which is a not a bad tool. It creates a searchable record, you can send a message at your convenience, the respondent can deal with it at their convenience, and you can easily send a message out to multiple parties. But unfortunately, since it's so easy, our e-mail boxes get clogged, and some people don't respond in a timely manner. I once asked a PM who reported to me, "What's a problem that you've gotten used to but that really should be fixed." Her response was that someone never responded to his e-mails.

She would e-mail him, then text him, then go talk to him. He was overwhelmed, and one of the ways he managed his workload was to ignore e-mails.

One way to manage project electronic communication is a separate tool like Basecamp or Slack. The advantage of the more modern tools is that they organize the history of communication much better than e-mail. How many times have you said, "Bill answered that question a while back, but I can't seem to find the e-mail," or needed to on-board a new team member by forwarding them an e-mail thread that was a mess. Getting a team who are used to communicating one way (e.g., e-mail) to adopt a new tool is a challenge that should not be undertaken lightly and which must have buy-in from top to bottom. Without vigilance, you'll find communication slipping back into the old familiar ways.

## Project Management (Individual)

The most popular project management software is Microsoft Project. It's a great tool for the PM to build a work breakdown structure (WBS), schedule, and a Gantt chart. It's not a good tool for communicating with team members or keeping the project on track.

When I worked at Microsoft on the Kinect game system, we had a prototype build coming up. We only had a few days reserved on the manufacturing line, and lots of components had to be built elsewhere and delivered to the factory for assembly on a certain date, or we would miss our window and throw the whole program into jeopardy. This situation is well suited to a waterfall-style plan, so I used MS Project to build a schedule and a Gantt chart, showing all the tasks that would need to be completed for the build to be successful. I showed it to the program managers, who acted like they had never seen a Gantt chart before. We discussed the critical path, agreed the schedule was tight but doable, and then they told me not to show the schedule to anyone. I instead was to build a spreadsheet with task, owner, and date needed, and that would be shared with the team because, even at Microsoft, no one liked Microsoft Project.

## Project Management (Team)

There are more modern web-based project management tools that allow you to build the WBS and schedule, make sure the people to do the work are available, communicate status, leave notes about tasks, and build status reports. These tools typically charge a subscription fee of ten to fifty dollars per person per month, and to be effective, every team member will need a license.

One that I like is LiquidPlanner, which is well designed for adaptive project management. LiquidPlanner's big advantage is the way it deals with how long a task will take: the duration is entered by the individual contributor as a range and can be easily updated by the people doing the work as their understanding of the task improves. The impact to the schedule is immediately calculated, and the PM can create a clear report that shows what tasks have changed and explain why.

If you're going to adopt a new tool, don't underestimate the level of effort needed to on-board the team. It was months of effort to get 80 percent of the team using LiquidPlanner effectively, which was better than I expected.

## What Every Individual Contributor Should Know About Adaptive Project Management

Project management in an innovative environment is the art of herding very smart cats. It's the project manager's job to make sure everyone is working on the correct tasks, that risks are being managed and communicated, and that the stakeholders (i.e., the decision makers) have everything they need to make informed decisions. It's not the PM's job to make decisions, but only to make sure the needed decisions are being made by the right people, who have all the information they need (or at least as much of it as possible). The PM needs certain things from the individual contributors (ICs). All of the advice can be summed up in one sentence:

> *Communicate quickly, completely, and candidly.*

If you just do that, you can't go wrong.

## Give Input on the Creation of the Plan

The PM is responsible for pulling together input from the team in order to create the plan by asking the following questions:

- What tasks are needed?
- What are the dependencies?
- How long will the tasks take?
- What assumptions are we making?
- What are the risks?

The team will be responsible for delivering to this plan, which is much more likely to go well if everyone in the team has given their input. It's critically important to capture all of the tasks that will be needed; otherwise, the team may not have all the skills needed to finish the project. Estimating how long a task will take is challenging, especially if it's something you've never done before. Realistic estimates are best; optimistic ones are the worst. It's the PM's job to communicate to the stakeholders how confident you are about the schedule.

If the plan doesn't meet the stakeholders' needs (e.g., a needed design mock-up won't be ready for a board meeting already scheduled), the team should gather to discuss how to adapt the plan, possibly by adding resources or shifting some scope to later in the program.

## Follow Your Company's Procedures

It might not sound important, but the difference between a successful project and a struggling one can be as simple as everyone filling out their time cards and other routine tasks. I once worked with an engineer who felt that filling out his time card by the deadline was beneath him. This made accurate status reports impossible, delayed invoices, and meant that basic metrics used to run the business, like "How many billable hours did we work last week?" were inaccurate.

## Keep Your PM in the Loop

It's the responsibility of the PM to be able to answer questions about their projects from the stakeholders, executive team, and functional managers. To be successful, they need to be in the loop.

If something goes wrong, let your PM know ASAP. I heard a story about two engineers in a small company who both screwed up. When the first screwed up, he didn't tell anyone. His manager, who used computer logs to figure out what had happened, discovered the mistake. When the engineer was asked about it, he lied. When shown the evidence, he admitted that it was his fault, but by then, it was too late, and he was rightly terminated. When the other engineer screwed up, she immediately told the same manager what went wrong and led the charge to fix the problem. Everyone saw her as a hero. We all screw up sometimes; the difference between a hero and a villain is how we respond to our mistakes.

Similarly, if you figure out that some of the inputs or assumptions in the plan are wrong, let your PM know right away. Maybe you've worked fifteen of the expected twenty hours on a task, and you now estimate it will take thirty hours to complete. Let your PM know immediately so the plan can be updated.

## Communications with Stakeholders

As part of the planning process, the PM will create a communication plan. Typically, the PM will be the primary point of contact with the project stakeholders, but sometimes, it makes sense for the

ICs to communicate directly with some of the stakeholders, especially if the discussions are highly technical. If that's the case, it's critical that the PM is in the loop for many reasons, including:

- In these conversations, scope increase is often requested (e.g., "While you're changing that code, why don't you add an option to change the units to metric"). Managing scope is one of the most difficult and critical tasks for a PM. Many a project has failed due to poorly managed scope creep.
- Information that invalidates an assumption might come up in the conversation.
- The IC might share information about a risk without informing the PM. When this risk isn't included in the status reports or the risk registry, the team loses credibility with the stakeholder.
- The PM needs to know the status of the project to complete reports and communicate with the internal team.

For e-mail correspondence, a cc may be sufficient. For a phone conversation, the IC should share meeting minutes with everyone directly involved in the conversation and the PM (or they can invite the PM to join the phone call and ask them to keep minutes).

## Communicate Quickly, Completely, and Candidly

As I said in the beginning, it all comes down to communication, but your PM can't follow every detail. Your PM doesn't need to know how you laid out a circuit board, but they do need to know why you had to violate design rules to make everything fit and why you think that's OK. Your PM doesn't need to know every line of code, but they do need to know why you're concerned that you might run out of memory. When in doubt, communicate more; if you're oversharing, your PM will let you know.

## What Every Stakeholder Should Know about Adaptive Project Management

What every stakeholder wants is an accurate and precise schedule with milestones that correspond to their needs and delivery dates that meet those needs. They want to pay a fixed cost that's as low as possible. They want the "real" work to start right away. They want to reap all of the rewards of a successful product but have none of the risk. If innovation is needed, they probably won't be getting all of these things.

### Schedule

Adaptive project management is all about managing the uncertainty that is fundamental to projects where innovation is needed. Just ask yourself, if no one in the world has ever done this and my team isn't quite sure how they're going to do it, how could an accurate schedule possibly be created?

This doesn't mean your milestones are unimportant. Communicate what you need and when you need it. Make clear what's **needed** and what's **desired**. Your project team should build a plan that maximizes the chance of success and communicates in advance whether or not your deliverables will be ready on time.

## Cost

Whether you have an external team or an internal one, you will probably have a budget to develop the project. But if innovation is needed, how much effort the project will take is unknown, so how do you manage a budget? Since we asked everyone to put away their optimistic hats while building the plan, you might be OK and even come in under budget (assuming no scope creep). In practice, even pessimistic estimates often end up being optimistic because one of the risks turned into an expensive issue that wasn't part of the estimate. I see three approaches to cost:

- Keep enough budget in reserve to deal with risks becoming issues, scope creep, or other drivers of budget woes.
- Have scope you can eliminate to reduce cost and shorten the schedule. These features need to be placed late in the development process.

It's no help to eliminate a feature that you've already developed.

- Have appropriately spaced stage-gate reviews and kill or descope the project as soon as you can see that the cost is beyond your means.

## Handle Disappointing News with Grace

You will hear bad news. The team will discover a problem that wasn't on the risk register; a key team member will leave to take a job elsewhere, and it will take a few weeks to bring a new person up to speed; a component that is normally available is out-of-stock. You've asked your development team to do something that hasn't been done before, so don't be too shocked if they didn't foresee every possible problem. If you handle bad news poorly, it will discourage the team from communicating honestly with you, and that will be far worse than whatever setback you were upset about.

## How to Run a Company That Can Do Adaptive Project Management

Long ago, in a city not far away, I worked for a very profitable company that got its start in the garage of one of its founders. We had grown to about 120 people, with sales made worldwide. Our products were the undisputed gold standard of the field. The employees were well paid and generally quite content. What's not to like?

But if you looked closer, there were problems. Our product line had never been refreshed, twelve-year-old designs were getting harder and harder to build as components went end-of-life, the manufacturability and reliability of our flagship product were poor, and new product development was sluggish and unfocused.

I joined the company as an engineer, but it was obvious that we didn't need our twentieth engineer; we needed our first project manager. As such, I wrote the company's first requirements document and got the stakeholders to agree to the product definition. I did what I could to add rigor to the development effort.

When I took over the project to write the software for the refresh of our flagship product, I created a giant flowchart that showed every possible interaction a user could have with the product. This flowchart acted as our requirements document; it accelerated the progress and increased the quality of our output.

Leadership tolerated what I did but never acknowledged the value in creating a development process. As a result, the hardware didn't meet the requirements, and a redesign project that should have taken less than three years to complete instead took seven. Sure, there were technical challenges, but they paled in comparison to the delays caused by the lack of project management and a product development process.

This is a common problem in growing companies. At first, communication is great because everyone knows what everyone else is doing. Priorities are obvious since there's only one project. But things start to break down as employees and projects are added. Just like people, healthy companies must grow and mature. They go through stages of development, and project management should grow along with the company. Too often, as companies grow, project management is one or two stages behind where it should be.

Here are four ways to grow project management as your company grows:

## 1. Have a Requirements Document.

Every project should have a requirements document that describes what the goals of the effort are and what "done" looks like. If something is required before the project can be done, it must be in the document. It can be short and simple or long and detailed, depending on the situation. This document should be approved by all of the stakeholders.

A key aspect of every requirement is that it must be verifiable; there can be no debate on whether you meet the requirement. *"It will be light"* is not a requirement, but *"It will be less than 500 grams"* is. It's better to say *"The product shall weigh less than TBD grams,"* since it makes it clear that you need to settle on a value quickly.

By putting this in writing, you can avoid false consensus, where everyone thinks they know what the end product will look like but they're wrong. You will also need a process to update this document because there should be changes as you progress, and all of the stakeholders should understand what they are and why the requirement is changing.

## 2. Have a Process to Start and Stop Projects.

As your organization grows, you'll need a process to green-light new projects, making sure you have a requirements document and the resources to do the work. You also need a way to kill the projects that don't make sense as soon as possible. Having a prioritized list of every project will help when there are resource conflicts. Finally, keep a list of pending projects so that good ideas have a place to wait until you have the resources to start the effort. Without an official place to put projects on hold, you'll be tempted to start projects that are a good idea even if you don't have the resources for it.

## 3. Treat Project Management like Your Other Disciplines.

Grow your company's project management maturity along with the size and number of projects that are happening. If your company is big enough to have a director of mechanical engineering, it's probably big enough for a director of project management who is responsible for mentoring the project managers and developing good processes.

You should also make sure you have top-quality tools. I've seen companies skimp here, and it just doesn't make sense. If you're paying your staff a good salary, a tool that increases everyone's productivity will have a positive ROI.

## 4. Your Processes Need to Be Just Right.

Proper process is critical to running a healthy company. If you just let everyone do what they want, a rogue trader may cost you two billion dollars. If you run a multinational corporation like a startup, there's no way for management to say, "We need to focus on the Internet" and make things happen. The key is to have the appropriate level of management such that it allows people with good ideas to bring value to their projects and to the company while still allowing management to set priorities and direction.

You need to guard against processes that don't add value. Update your old processes to make sure they fit the reality of the company that is and will be, not the company that was. As painful as too little process is, it's less painful than too much process, so when in doubt, keep process to a minimum.

## Building an Innovative Organization

The culture of innovation goes beyond how you run projects and should influence the entire company, from how you hire staff to how you do performance reviews and manage time off. There are many modern management paradigms that work well with adaptive project management, but it's nearly impossible to manage a project with service leadership and risk-taking if the executives are screaming at the employees.

A holacracy is a system of organizational governance in which authority and decision-making are distributed throughout a holarchy of self-organizing teams. In a holacracy, *predict-and-control* is replaced with *sense-and-respond*, which is the core idea of adaptive project management. There is a hierarchy of circles (a self-organizing team), with each one receiving purpose and accountability from the circle above. This is similar to the "assign problems to solve, not tasks to carry out" aspect of adaptive project management.

The book *Reinventing Organizations: An Illustrated Invitation to Join the Conversation on Next-Stage Organizations* by Frederic Laloux talks about building organizations based on trust and self-governance. Adaptive project management would be comfortably at home in an evolutionary organization where teams are self-led.

But adaptive project management can even thrive in more old-fashioned organizations as long as the leadership of the organization accepts this approach. That means they can't scream at people when things fall behind, but rather will ask why and listen to the answer.

## Now, Go Forth and Innovate

I think back to the best-run project I ever worked on. The team worked hard, creatively, and collaboratively. The design was exactly what the stakeholders asked for. I'd suggest you go out and buy this great product, but the marketing department killed it right before we released to manufacturing because the market had changed. The team was crestfallen. What I said was, "We did a great job. We learned a lot. We were paid a great salary. It would have been great to see our work out in the world, but we should be no less proud of what we did."

So, success is not guaranteed. Good project management won't change the laws of physics, the regulations of the US government, what products your competitors launch, or any of the hosts of reasons projects and products fail. But it can eliminate failures for reasons that are within your control, increase how much the team learns from the process, and make the process one that draws people to work with you.

And remember that innovation is hard. You're solving a problem that hasn't been solved before. You can't just type a destination into Google Maps and follow a well-worn route. You're Lewis & Clark, Lindbergh & Earhart, Gagarin & Armstrong. Enjoy the journey.